YOUR KNOWLEDGE HAS VALUE

- We will publish your bachelor's and master's thesis, essays and papers

- Your own eBook and book - sold worldwide in all relevant shops

- Earn money with each sale

Upload your text at www.GRIN.com and publish for free

Bibliographic information published by the German National Library:

The German National Library lists this publication in the National Bibliography; detailed bibliographic data are available on the Internet at http://dnb.dnb.de .

Imprint:

Copyright © 2018 GRIN Verlag
Print and binding: Books on Demand GmbH, Norderstedt Germany
ISBN: 9783668741133

This book at GRIN:

https://www.grin.com/document/431375

Patrick Kimuyu

Learning As Humans Survival Adaptation

GRIN Verlag

Learning As Humans Survival Adaptation

Name: Patrick Kimuyu

Contents

Introduction

Psychology of learning consists of two concepts: education and psychology. Educational psychology, therefore, is the application of pure science in education with the purpose of modifying and socializing the behavior of an individual. According to Mintz (2013), educational psychology is a description and explanation of the learning experience of a person from their birth to old age. It is that psychology branch dealing with learning and teaching. Krishna (2004) states that, the meaning of educational has been defined differently, but the notable definitions are by Stephen who defines educational psychology as the "systematic study of the educational growth and development of a child" (p. 22). On the other hand, Judd (2009) describes educational psychology as "the science that explains the changes that take place in the individuals as they pass through the various stages of development" (p. 11). In short, this is the science of learning.

This psychological branch is among the applied psychology branches, which is concerned with applying techniques, principles, and other psychological resources to the solutions of the confronted problems by the teacher who is trying to direct the children's growth towards a set of defined objectives. More distinctively, we can say that this psychology attempts to understand:

- Children, their development, needs and potentials
- The learning situation, which includes group dynamics and their effect on learning
- The learning process, nature and means of ensuring that it is effective

Learning psychology covers five major areas: the learner, learning process, learning situation, teaching situation, evaluation of learning performance, and the teacher. Psychology of learning has important methods of studying it, which are essentially general psychology methods. The methods used to study psychology of learning include the introspection method, observation method and experimental method (Tyagi, 2011).

Learning

In psychology, learning is a process leading to a relatively permanent change in the potential behavior due to experience or practice. It is also the process of attaining modifications in the existing skills, knowledge, tendencies, and habits through practice, exercise or experience. Learning emphasizes four attributes, which are:

- As a process, learning brings about a lasting behavior change.
- Learning does not also change due to fatigue, illness, use of intoxicants, and maturation.
- Learning manifests in individual activities, therefore, it is cannot be directly observed.
- Learning is dependent of experience and practice.

Learning has definite characteristics. Mc Law (2013) identifies the following characteristics:

i. Learning happens throughout the life since it is a constant behavior modification.
ii. Learning reaches the human life in all aspects. Therefore, it is described to be pervasive.
iii. Learning is an engagement of the whole person, emotionally, socially, and intellectually.
iv. Time is among the elements of learning. It is a developmental process.
v. Learning often results to a behavior change.
vi. Learning is receptive to incentives. Usually, positive incentives, for example, rewards are motivational and effective to learning as compared to negative incentives like punishment.
vii. Learning and interest are positively related. An individual will be interested to learn the things he is attracted and interested. This is especially among sports such as most footballers are likely to learn playing football rather than adding fractions.
viii. Goals are of utmost concern in learning. The goals are articulated as observable behavior.
ix. Learning is dependent on motivation and maturation.

Types of Learning

Learning can be described in many ways. It can be formal, informal, and non-formal. Formal learning is organized and intentional. This form of learning occurs in a formal educational institution. Non-formal education is also organized and intentional. However, it is flexible as compared to formal learning. Informal is very different because it is incidental. This form of learning occurs throughout an individual's life and it is never planned. Learning can also be classified as individual or group learning. It can be referred in these two dimensions depending upon the number of persons that are involved in the learning process (Mintz, 2013).

The other classification is dependent on the activity that is involved in learning. This classification includes motor learning where learning is specifically in the use of muscles such as walking and discrimination learning, which entails the discrimination act such as discrimination by an infant of milk and water. It also involves verbal learning where use of words is involved in learning; concept learning when concept formation is involved in learning; sensory learning, which entails the use of sense and perception for learning (Kirk, Macdonald & O'Sullivan, 2006).

Learning Theories

Learning is a process focusing on the happenings of learning to take place. These explanations of the happening are the learning theories. The learning theories, therefore, attempt to give a description of the process of learning in either animals or people, which is important in helping in understanding the intrinsic complex learning process. The theories according to Mintz (2013) have two identifiable values. They help in providing a conceptual framework and vocabulary used to interpret the learning examples observed and besides this, they are important in suggesting the places to look for solutions to learning practical problems.

Learning theories fall under three main philosophical frameworks, which are cognitive, behavioral and constructivism. Behaviorism focuses on the learning aspects that are objectively observable. Cognitive theories explain the brain-based learning. Additionally, constructivism examines learning as a process by which ideas and concepts are actively constructed by the learner (Dean & Jolly, 2012).

Cognitive theories cite that different learners have distinct styles leading to learning influence. It is believed that the most crucial thing is that the student has a promotion and prevention focus. When a student has a prevention focus, the negative outcomes are sensitive, the student seeks to avoid errors, and security concerns are the student's drive. Promotion focus is more of the positive outcomes by the student. Once there is a fit in these two elements, learning is enhanced (Mintz, 2013).

Behavioral theories are discussed in two categories: the Stimulus Response (S-R) theory with reinforcement and Stimulus-Response theory with reinforcement.

Stimulus-Response Theory with Reinforcement

Trial and Error Theory by E.L Thorndike

The first psychologist to forward this theory was Edward Lee Thorndike. According to Lee, learning occurs because of the stimulus and response bond. Further, he identifies that learning occurs through an approximation and correction process. An individual makes different trials with some unsatisfactory responses but the individual makes more trials until the satisfactory responses are achieved (Law, 2013).

Based on the trial and error theory, Thorndike provided certain learning laws. The first element of this theory is the law of readiness, which explains why learning only occurs in cases where the learner is willing to learn. In the absence of willingness, there is no effort that can be used when a child is not willing to learn (Dean & Jolly, 2012). This law relies on the dictum that states 'you can lead a horse to the pond, but you cannot make it drink water unless it feels thirsty.' Precisely, when a child is willing to learn, he/she gets information more quickly, with great satisfaction, and effectively than when a child is not prepared to learn. This law of readiness, therefore, denotes that mental preparation is important for action. Learning failures occur when one forces a child to learn something when he/ she is not ready.

Educationally, this law draws the teacher's attention to the child's motivation. The teacher has to consider the student's psychobiological readiness to make sure that the learning experiences are successful. Learning/curriculum experiences should be in accordance with the maturity's mental level of the child. Failure to this, poor understanding occurs and readiness to learn will vanish (Dean & Jolly, 2012; Mintz, 2013).

Second is the law of exercise. This law clarifies on the practice role in learning. As this law cites, learning is efficient when there is exercise or practice. This law rhyme with the dictum 'practice makes perfect.' Nonetheless, this law is split into the law of use and disuse. The law of use means that the stimulus and response connection is made stronger by its exercise, occurrence or its use. The law of disuse, on the other hand, means that failure of the stimulus and response modifiable connection over a certain time leads to decreased strength of the connection. This signifies that any action that neither is nor practiced for a certain period decays (Law, 2013).

Educationally, exercise is important in learning. The teacher has the responsibility of repeating; giving sufficient practice in some subjects that will help fix the material in the students' heads. Thorndike revised this law and identified that; law of exercise is important but not sufficient in learning. Practice should always be followed by reward or satisfaction by the learner. Therefore, learners require motivation, in order to participate in the learning process (Law, 2013).

Third is the law of effect. This is the most important among Thorndike's laws. It states that when stimulus and response connection goes along with satisfaction, the connection is strong. The law indicates that responses producing discomfort to a learner are weakened while those responses producing satisfaction are strengthened (Law, 2013).

In learning, this law implies the need for feedback. Therefore, a satisfaction consequence must be associated with learning trials. The teacher can use rewards to help strengthen some responses while punishment is used to weaken others. However, in school learning, use of rewards is more advantageous than the use of punishment. The teacher can exploit reward system as a motivation to students (Dean & Jolly, 2012).

Operant conditioning by BF Skinner

This is a punishment and reward for behavior in learning. An association is established between a consequence and a behavior. Skinner recognized three operant or responses that follow a behavior. The first aspect is the neutral operant, which is environmental operant that can neither enhance nor reduce the repetition of the behavior, whereas the second aspect is the reinforcement, which strengthens the following behavior. They can be either positive or negative. Reinforces, which good events are presented after a behavior are positive while those that involve removing a bad event after a behavior display are negative. On the other hand, the third response involves punishment, which is considered as an unpleasant outcome that leads to a decrease in behavior. This theory has different implications on learning. The theory is meant to condition study behavior. Teaching accelerates learning, therefore, for effective teaching; the teacher should have a good arrangement of reinforcement contingencies. The teacher can use incentives to students is a good way of reinforcing student behavior. During the learning process, a student can obtain the undesirable behavior. This unlikeable behavior is conditioned to the subject, teacher, or the classroom. Good behavioral contingencies, acceptance, atmosphere, esteem and affection help the child in the teacher and subject approach. In instances where the child is not serious with the study, use of negative reinforcement like criticism can be used but in cases of seriousness in the study, the teacher can use positive reinforcement such as rewards (Law, 2013).

The theory can also be used to manage the problem in behavior. Both undesired and problematic behaviors are seen in class. Operant conditioning helps shape the behavior of the students. McCown and Snowman (2011) state that operant conditioning helps in behavior modification through "ignoring undesirable responses and reinforcing desirable responses" (p. 224). Therefore, a teacher should not admit negative contingencies such as punishment rather than adopting positive contingencies including encouragement.

Classical Conditioning by Pavlov

This learning acquired through experience. The theory identifies that for one to embark on a new task; one should master and practice the present task effectively. The same case applies to learning where the student must respond to particular information before he/she can be embarking in a new one. Teachers have a task of identifying the methods of student motivation to help them learn. They should be aggressive with different strategies that enhance teaching and learning (Law, 2013).

Classical conditioning helps learn most emotional responses. A positive or negative response comes through the paired stimulus. For instance, provision of the essential materials for school helps the student develop a good feeling about learning and school while; on the other hand, punishment discourages attendance of school (Law, 2013).

Factors Affecting Learning

Learning as identified earlier is the process through which, knowledge, attitudes, concepts and skills are understood, acquired, and applied. Every person engages in learning either as a conscious process, sub-consciously, or unconsciously. Learning equips individuals with the ability and competence to enhance their functioning in their environment. As we learn through teaching and instruction, we also learn through the experiences and feeling. Experiences and feeling, being touchable part of our lives influence why, how and what we learn (Hanrahan, 1998).

Learning can be partly social or affective and partly cognitive. It is a cognitive process because it includes functions of perception, attention, reasoning, drawing conclusion and giving significance to observed phenomena. These are all mental processes of an individual relating to intellectual functions. Cultural and societal context, which individual function, experiences and feelings influences the concepts, ideas, images and comprehension of the world. This explains the social and affective process of learning (Hanrahan, 1998).

Our ideas, knowledge, beliefs, concepts, attitudes and skills that individuals acquire are as a result of these merged processes. The learning process involves feeling, experience, and cognition. Different people differ greatly in respect to learning capacity, ability, and interests. For example, some people are good in gaining practical skills such as electrical repairs. There are different factors that affect learning leading to differences in what individuals learn and how they learn (Grabowski & Jonassen, 2012).

Maturation as a Learning Factor

Maturation is one of the significant factors affecting learning, and it is defined as a continuous growth occurring within an array of environmental condition. In theory, maturity occurs regularly to an individual without stimulation conditions such as practice and training. Learning possibility is only attainable when a certain maturation stage is reached. Training and exercise are also is only fruitful when an individual achieves a certain maturation stage. This determines how ready the child is for learning and this is the determinant for the readiness of learning in a child. There are individual maturation differences meaning that there is a variance between different people. Similarly, the learning capacity differs to people in the same age level, which is attributed to the different maturation levels. This brings about specific skills that are learned by children maturing earlier than others do (Dean & Jolly, 2012).

The 3R's (reading, reckoning and writing) is only possible after brain and muscular capacities maturation. The learning rate is mainly dependent on cerebral cortex maturation. When cortical tissues deteriorate because of old age, the learning ability also declines. Learning can only occur when a certain age of maturation is reached. Practice is significant when it goes hand in hand with maturation level. Thus, it is essential for the teachers to identify their pupil's maturation level (Dean & Jolly, 2012).

Attention and Perception

Attention is a present conscious in our lives and common to all mental activity types. All the activities of an individual are based on attention and interest. Achievement of goals is only successful when the attention of an individual is expressed to learning. According to Hanrahan (1998), attention entails the act of selective consciousness. Attention can be voluntary, selective or involuntary. Voluntary attention is when an individual only looks for the information that is of personal relevance. Selective attention is when an individual only chooses the relevant information. Involuntary attention occurs when something is exposed to a person in an unexpected manner.

Despite these differences in forms of attention, they all possess the same characteristics. Attention entails focusing on only one object with all the other objects being the attention's margin. Attention is also selective since one object is preferred to the others. Nonetheless, it constantly shifts from the focus to the margin, and it is divided between two tasks of the mind (Robinson, 2013).

Attention is important in learning because; every moment of an individual is drawn to different stimuli in the environment. Attention is the cause for the concentration to one single object in the environment. Attention helps the students clear all the other vivid objects and, therefore, improve interest, efficiency, motivation, and readiness of the learners (Robinson, 2013).

On the other hand, perception is the process of exposure to information, attending to this information, and its comprehension. Perception is a mental process through which knowledge from the environment is integrated. An individual receives the information using the sense organs and some of this information is selected and organized to give meaning. In short, it is the sensation plus meaning (Hanrahan, 1998).

Learning is dependent on what the individual percepts. The possibility of an individual to perceive correctly leads to proper learning. Correct percepts lead to the proper direction of learning. Sensations give the first impression leading to the appropriate perception, which further leads to the understanding of an idea, object or experience. Learning is dependent on efficient and accurate perception and similarly perception is dependent on sensation. Hanrahan (1998) reports "quantitative studies have found strong relationships between student perceptions of the classroom goal orientation and their own use of deep learning strategies" (p. 48).

Motivation Factor of Learning

Motivation is the inspiration propelling somebody to a certain action. It is an internal condition, which activates and shows the direction of feelings, thought, and action. It is a process where the internal energies of a learner are directed towards the achievement of a certain goal. Motivation may occur in one task while, in another, the individual can be unmotivated. When motivated, an individual can work tirelessly to achieve their goals. Motivation according to Abraham Maslow leads to growth and satisfaction is a significant factor in motivation. It is crucial for the teacher to identify the basic needs of students and attend to them to their importance level (Dutton, Dutton & Perry, 2002). Before teaching students, the teacher should first think of the health of the learners.   Praise of students by the teacher is a certain study leads to an effort to maintain that level. Students also need respect and recognition in classes as well as their views. This is meant to boost and develop the students' confidence (Mintz, 2013).

Motivation is also boosted once the teacher makes the outcome of the lesson known to his/her students. This knowledge builds attention by the students. Feedback is important to sustain the student interests in the classroom. The teacher has the task of letting the students know their performance in learning (Mintz, 2013).

Conclusion

Conclusively, this psychological research is important in learning and it is significant in promoting learning by the student and improved teaching. The learning theories identify the important factors that enhance the learning and specifically identifying the effect of rewards and punishment. The theories give the teachers and learners the methods to improve learning through the understanding of the complex learning process. It provides a framework for interpreting learning examples that are observed. Nonetheless, the theories are suggestions for solutions for the practical problems in education. These theories are a direction to the variables used in finding solutions. The factors affecting learning responds to the why an individual performs some tasks better that others at the same level.

References

Dean, K. L., & Jolly, J. P. (2012). Student Identity, Disengagement, and Learning. *Academy of Management Learning & Education, 11*(2), 228–243.

Dutton, J., Dutton, M & Perry, J. (2002). How do Online Students Differ from Lecture Students? *Jalan,* 6(1), 1-20. Retrieved from http://uwf.edu/ATC/Guide/PDFs/how_online_students_differ.pdf

Grabowski, B. L., & Jonassen, D. H. (2012). *Handbook of Individual Differences Learning and Instruction.* New York, NY: Routledge.

Hanrahan, M. (1998). The Effect of Learning Environment Factors on Students' Motivation and Learning. *International Journal of Science Education, 20*(6), 737-753. Retrieved from http://eprints.qut.edu.au/1352/1/hanrahan_ijse.pdf

Judd, C. A. (2009). *Essentials of Educational Psychology, 2E.* Uttar Pradesh, India: Vikas Publishing House Pvt Ltd.

Kirk, D., Macdonald, D., & O'Sullivan, M. (2006). *Handbook of Physical Education.* Thousand Oaks, CA: Sage Publishers.

Krishna, V. V. (2004). *School Psychology.* New Delhi, India: Discovery Publishing House.

Law, H. (2013). *The Psychology of Coaching, Mentoring and Learning.* Hoboken, NJ: John Wiley & Sons.

McCown, R., & Snowman, J. (2011). *Psychology Applied to Teaching.* Stamford, CT: Cengage Learning.

Mintz, S. (2013). *The Psychology of Learning and the Art of Teaching.* Columbia, Columbia University Graduate School of Arts & Sciences Teaching Centre.

Robinson, S. (2013). Student Response to Risk in Classroom Learning Games. *Academy of Educational Leadership Journal, 17*(4), 1-12.

Tyagi, A. K. (2011). *Psychology of Teaching, Learning and Process.* Oxford, UK; Pinnacle Technology.

YOUR KNOWLEDGE HAS VALUE

- We will publish your bachelor's and
  master's thesis, essays and papers

- Your own eBook and book -
  sold worldwide in all relevant shops

- Earn money with each sale

Upload your text at www.GRIN.com
and publish for free